Judith S. Seixas

Alcohol -
What It Is,
What It Does

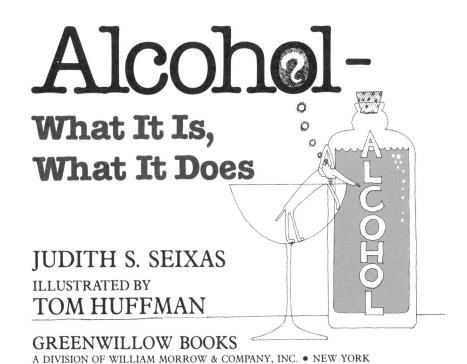

JUDITH S. SEIXAS

ILLUSTRATED BY

TOM HUFFMAN

GREENWILLOW BOOKS
A DIVISION OF WILLIAM MORROW & COMPANY, INC. • NEW YORK

Greenwillow
Read-alone

First Edition
26 25 24 23 22 21 20 19 18 17

Library of Congress Cataloging in Publication Data
Seixas, Judith S   Alcohol—what it is, what it does.
(Greenwillow read-alone series)
Summary: An easy-to-read introduction to the facts
about alcohol, what it is, where it can be
found, and its effect on the mind and body.
1. Alcoholism—Juvenile literature. [1. Alcohol.
2. Alcoholism] I. Huffman, Tom. II. Title.
RC565.S42   613.8'1   76-43344
ISBN 0-688-80080-7   ISBN 0-688-84080-9 lib. bdg.

To Peter,
Abby,
Noah   – and for Frank

Contents

Introduction

If there were a glass of something
on a table and you were really thirsty,
would you drink it before you knew
what was in it?
You've heard people say,
"Drink your milk, it's good for you."
And you may have heard someone say,
"Don't drink so much soda,
your teeth will fall out."

But what do you know about alcohol
and drinks with alcohol in them?
You may have tasted some drinks
with alcohol in them.
And you may have seen drunk people.
Here's a book about alcohol
and drinks that have alcohol in them.
It is also about what happens
to your body and your mind
when you drink alcohol.
And it is about what happens to people
when they drink too much.

[1] Drinking

If you pick up a glass

of milk or orange juice,

you know that it is good for you.

It helps your bones get strong

and also helps you grow.

It is full of vitamins and minerals.

But a drink of alcohol is different.

It has lots of calories,

so it is fattening.

Yet it has nothing in it

that the body needs.

Then why do people drink alcohol?
Here are some of the reasons:

- to feel happy
- to feel grown up
- to feel less thirsty
- to forget troubles
- to do what friends do
- to enjoy the taste

Most people feel good
when they drink a little.
They also feel happy
and friendly and good
about other people.

But most people feel drunk

when they drink a lot.

When they get drunk,

they have trouble talking,

and their words get jumbled.

They feel dizzy.

They can't walk straight.

They might bump into a wall.

They might fall down,

or they might go to sleep.

Some people decide

they will not drink alcohol.

Their reasons are:

- they don't like the taste
- drinking makes them feel bad
- drinking gives them a headache
- drinking makes them vomit
- they believe drinking is a sin
- they don't take drugs of any kind

[2] Facts About Alcohol

Alcohol is always a liquid.

It comes in a can or a bottle.

It is never a pill or a powder.

It can't be "freeze dried."

There is no alcohol
in tomato juice,
grape juice,
soft drinks, sodas,
or vinegar.

Most of the alcohol that people drink
is in beer, wine, or hard liquor
such as whiskey, gin, vodka, or rum.
The law says that bottles of wine
or hard liquor must be marked
with either percent (%) or "Proof."
This shows how much alcohol
the bottle contains.

"100 Proof" means that

one half or 50%

of what is in the bottle

is alcohol.

The rest is water and flavors.

A can of beer,

a glass of wine,

and a shot glass of whiskey

hold different amounts of liquid.

But the amount of alcohol in each

is the same.

Most mouthwashes
and most cough medicines
have alcohol in them.
Anything that says "extract"
or "tincture" on the bottle
has alcohol in it.

TINCTURE

Look at the "Vanilla Extract"
in your kitchen.
It's marked "35% Alcohol."

VANILLA
EXTRACT
35%

Look at the mouthwash
or the cough medicine
in your bathroom cabinet.
What % alcohol is it?

MOUTH
WASH
?%
ALCOHOL

18

When you see a bottle marked

RUBBING ALCOHOL

don't drink it!

This is a different kind of alcohol.

It's poison!

Even a little will make you very sick.

It is meant to be used

on your skin to clean it,

or to cool you off

if you have a fever.

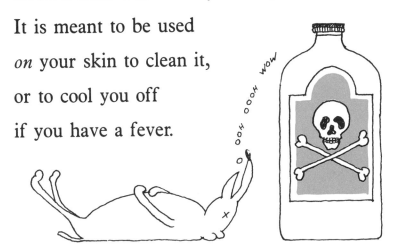

[3] The Body and Alcohol

There is no part of the body
that is not affected by alcohol
if it is taken in large amounts
over a long period of time.

Alcohol is not digested
like other foods.
As you swallow it,
it is absorbed through the lining
of your mouth
of your throat
of your stomach
of your intestines.

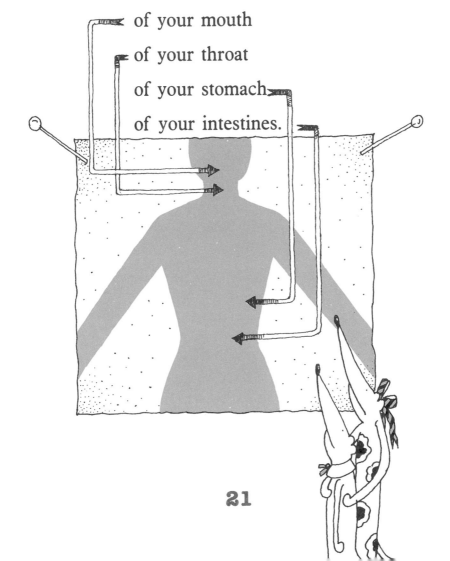

Alcohol gets into your bloodstream
and around your body very quickly.
Most of the alcohol a person drinks
is used up by the liver.
And so it is the liver
that is most often damaged
by too much drinking.

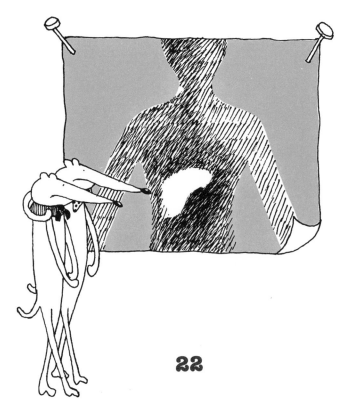

The alcohol that is not used up
by the liver
does not stay in the body.

One of the ways
alcohol comes out of the body
is in your breath.
Have you ever smelled
the breath of a person
who has been drinking alcohol?

It also comes out in sweat and in urine.

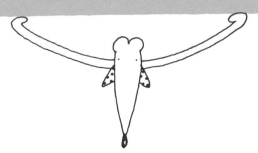

The less a person weighs,
the more alcohol gets
to every part of the body.
A very small person
who drinks
a mouthful of alcohol
feels the same as
a very large person
who drinks a cupful.

25

Alcohol gets to the brain
just a few minutes
after it is swallowed,
and it quickly changes
how the drinker feels.

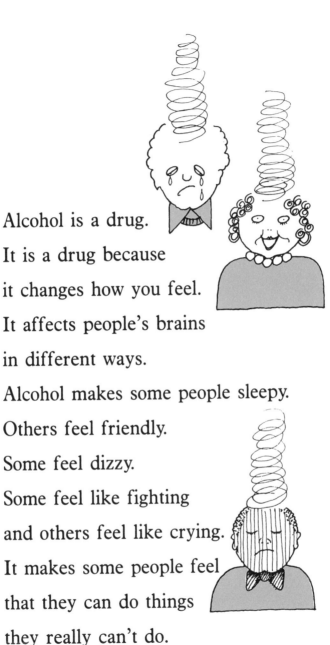

Alcohol is a drug.

It is a drug because

it changes how you feel.

It affects people's brains

in different ways.

Alcohol makes some people sleepy.

Others feel friendly.

Some feel dizzy.

Some feel like fighting

and others feel like crying.

It makes some people feel

that they can do things

they really can't do.

When people drink too much,
there is only one way
to get sober or "undrunk."
That is to let *time* go by.
The best way is to sleep.
In spite of what people like to think,
it is not possible to sober up
by drinking a cup of coffee
or taking a cold shower
or running around the block.

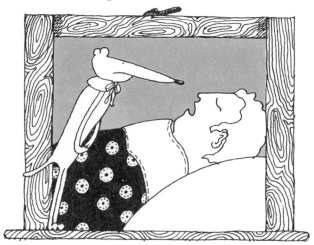

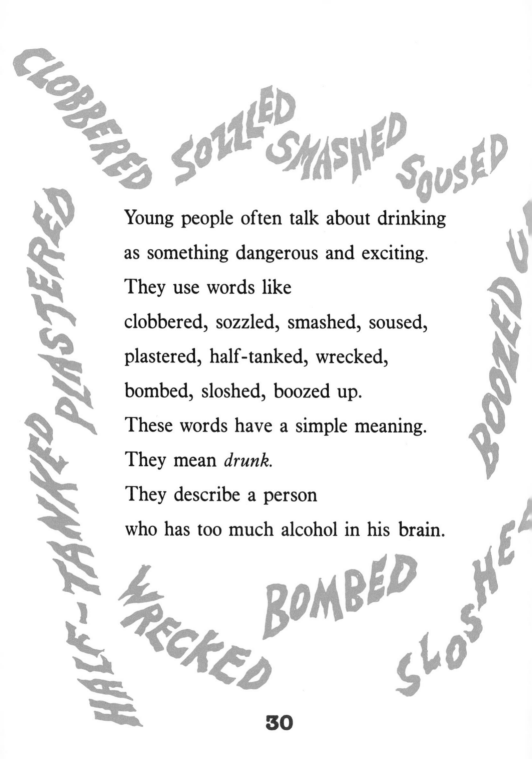

CLOBBERED SOZZLED SMASHED SOUSED PLASTERED HALF-TANKED WRECKED BOMBED SLOSHED BOOZED UP

Young people often talk about drinking
as something dangerous and exciting.
They use words like
clobbered, sozzled, smashed, soused,
plastered, half-tanked, wrecked,
bombed, sloshed, boozed up.
These words have a simple meaning.
They mean *drunk*.
They describe a person
who has too much alcohol in his brain.

Many people feel sick
the day after
they drink too much.
This awful feeling
is called a "hangover."

The best way
not to get a hangover
is not to drink too much.

In fact,
drinking anything very fast
or in big amounts
can make people feel sick.
But because alcohol is a drug,
drinks with alcohol in them
make people feel even sicker.

People can die from
an *overdose* of alcohol.
An overdose occurs
when a person drinks
too *much* too *fast.*
Some people make bets
about how fast they can drink ...
and they try
    "chugalugging."
They can end up in the hospital.

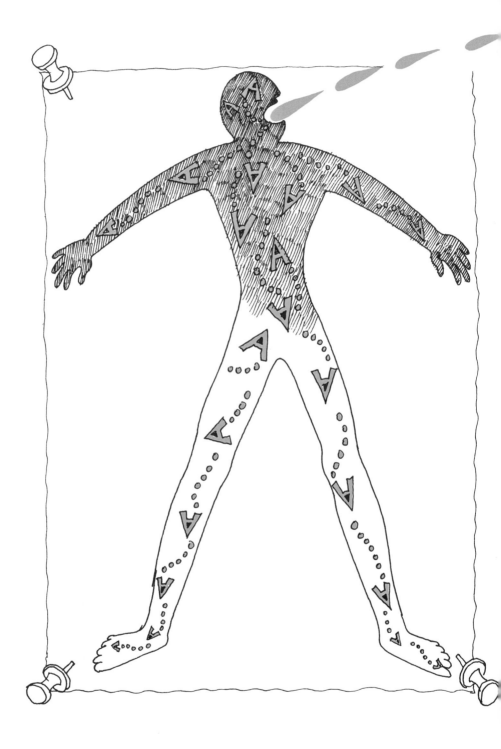

[4] Alcoholism Is a Sickness

People who drink too much
for years and years become alcoholic.
They can't stop drinking
any more than you can stop sneezing
when you're getting a cold.
Their bodies get used to having alcohol.
They feel sick and shaky and horrible
when they stop drinking.
Once a person is "hooked" on alcohol,
he needs help to stop drinking.

Alcoholism is a sickness
just as measles is.
But it is not catching
because there is no such thing
as an alcoholism germ.
But alcoholism does run in families.
Scientists don't yet know
whether alcoholism is inherited,
in the way small feet and blond hair are,
or whether it is a way of drinking
learned by watching others.

If you know a number of people who drink, there is a good chance that one of them is an alcoholic.

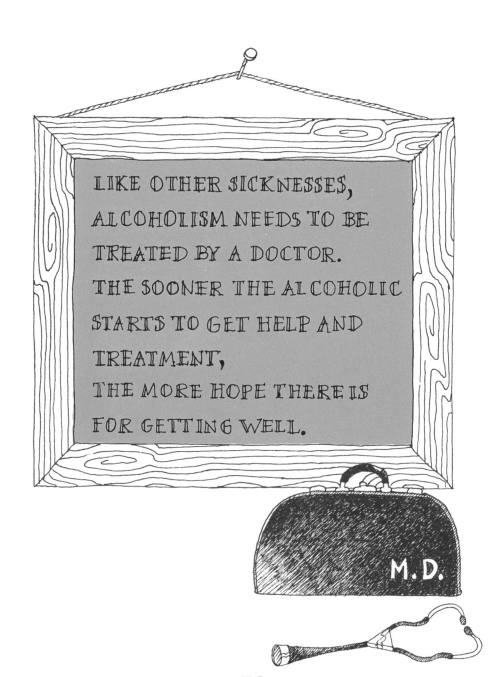

LIKE OTHER SICKNESSES,
ALCOHOLISM NEEDS TO BE
TREATED BY A DOCTOR.
THE SOONER THE ALCOHOLIC
STARTS TO GET HELP AND
TREATMENT,
THE MORE HOPE THERE IS
FOR GETTING WELL.

M.D.

Most alcoholic people
have jobs and homes and children.
They try very hard
to keep their families together
and to keep their jobs.
But drinking often gets in the way.

When someone close to you

is drinking too much,

you may feel scared or helpless.

Or you may feel that you want

to make the person stop drinking.

But you really can't stop another person

from drinking.

Alcohol affects people's ways of thinking.

That's why drunk people seem so crazy.

And that's why alcoholic people

are so hard to understand.

It is scary to be with a person who is drunk, because you don't know what can happen.

Some drunk people
fight and
throw things.

Other drunk people go to sleep.

You can help yourself

if you try to find something to do

that takes you away

from the drinking person

and gives you something else

to think about

until that person is sober again.

If there is an alcoholic
in your family,
you should know that
the alcoholic is not drinking
because of you.

IT IS NOT YOUR FAULT!

It is not your fault
if he drinks too much.
He is sick.
If you see a drunk person
lying on the street,
he needs the care of a doctor too.
He is also sick.
Like other sick people
many alcoholics do get well.

[5] You Will Have a Choice

Have you ever heard a person say,
"Let's have a drink"?
They usually are talking about
something with alcohol in it.

When someone asks you
what you want to drink,
they're probably not
talking about alcohol
but they are asking about a drink
that could go with your meal
or stop you from being thirsty.
But the time will come
when you'll be asked
if you want a drink . . .
a drink with alcohol in it.
Now you know something about
alcoholic drinks and what they do
to your mind and your body
and how they can affect
the people around you.

You know alcohol can make you feel good
and that it can make you sick.
You can choose to drink or not to drink.
Many people decide
not to drink alcohol at all.
Some people choose to drink
once in a while.

It's good to know
that you have a choice.
When the time comes,
you'll do what's right for you.

More to Talk About

1. Alcohol makes you warm.

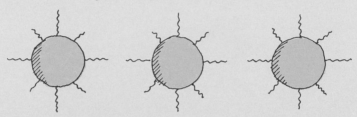

2. Beer is not alcohol.

3. Since vodka does not smell,
   it doesn't make you drunk.

No. It lowers your temperature.

Yes, it is. Most beer is 5% alcohol.

Yes, it does.

4. People drive better when they drink a little.

5. Drinking alcohol is a way to solve problems.

6. Coca-Cola and aspirin mixed together
make you drunk.

No. They *feel* as if they can drive better.

No. It often makes just one more problem.

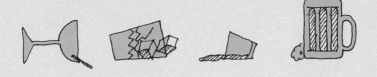

No. Alcohol is the drink that makes people drunk.

7. An alcoholic can stop drinking by using willpower.

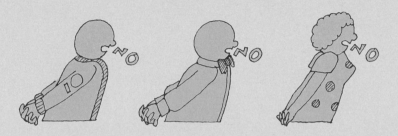

8. Every drink kills off some brain cells.

9. Rich people don't become alcoholic.

Willpower alone does not usually cure sicknesses.

No. There is no proof of that now.

Yes they do. Alcoholism hits rich/poor, black/white, smart/dumb, young/old, men/women . . . all alike.

10. Alcohol is a pain-killer.

11. There is alcohol in root beer.

12. Most alcoholics are "bums."

Yes. It is closely related to ether.

In big doses it puts you to sleep too.

No. Root beer is a soft drink.

There is no alcohol in it.

No. Only a few are found lying on the street.

Most are in their homes trying to keep

themselves and their families together.

⭐ There is help for alcoholic people in Alcoholics Anonymous.

⭐ There is help for their families and friends in Alanon.

⭐ There is help for the children of alcoholics in Alateen.